THE AUTHORIZED COLLECTION OF

HOLUSION ART

HOW AND WHY IT WORKS

CREDITS

NVISION GRAFIX, INC. — MICHAEL S. BIELINSKI & PAUL G. HERBER
HOLUSION™ ART — NVISION GRAFIX, INC.
ILLUSTRATIONS — NVISION GRAFIX, INC.
EDITOR — CHRIS TUCKER
PHOTOGRAPHY — STEVE CHENN
CREATIVE DIRECTION & DESIGN — STRICKLAND DESIGN INCORPORATED

U.S. PATENT NO. 4,135,502
FIRST PRINTING APRIL 1, 1994

PRINTED IN U.S.A.
ISBN NUMBER 0-9640923-0-1

NVISION GRAFIX, INC.
222 W. LAS COLINAS BLVD., SUITE 1840
IRVING, TEXAS 75039
U.S.A.

To Dad,
Merry Christmas '97
Love, Tasia, Ray, Hershey, Goof

Ho-lu-sion \hä-'lü-zhən\ 1. a trademark for an apparently meaningless pattern capable of fooling the human mind into seeing a three-dimensional image that does not exist. 2. a trademark for artwork containing a pattern capable of such mental trickery.

How to see the art:

1. First, remove the sticker from the dust jacket, releasing the transparent sheet. For better viewing of the images, place this transparent sheet on each work of Holusion Art. Then look at your reflection and watch a whole new world open up.

Or:

2. With the art almost touching your nose, just stare as if you were trying to look right through the page. Then move the page slowly away from your face. When the art is at a comfortable viewing distance, the image should come into focus. If it doesn't, don't worry. It may take several attempts before you see it.

Note: Don't be in too much of a hurry. Some people are unable to see the art because the moment they start to see part of the image, they rivet their attention on the page. Remember, stay relaxed and keep looking through the page until the whole picture appears.

Cartoonist Greg Evans, creator of the comic strip ***Luann,*** captured in four short frames what millions of people across the world have experienced with wonder, disbelief and frustration – Holusion Art by NVision Grafix. This comic message marked a symbolic milestone for the young founders of Holusion Art, Michael Bielinski and Paul Herber. Holusion Art, their creation, had become a part of American popular culture.

The concept for Holusion Art started in Dallas in 1992 as a collaboration between Bielinski and Herber. Herber, who had been associated with the B-2 Stealth Bomber project, approached Bielinski with some three-dimensional sketches he had done of the plane.

By coincidence, Bielinski had also been dabbling with 3-D images. After a brainstorming session and several months of late-night hacking, the first print of the Stealth Bomber rolled off the press in July of 1992. Herber sold several copies to friends in defense firms in California. When people began to glimpse the famous aircraft slicing through a sea of squiggles, it caused such excitement and disruption that the print was banned from the workplace. Holusion Art was born.

From its dual birthplaces of Dallas and Los Angeles, Holusion Art spread first across the United States, and then the globe. Two years after that first print appeared, Holusion Art is sold in more than fifty countries around the world, on all seven continents, including Antarctica.

Transcending the boundaries of language and nationality, Holusion Art has become a worldwide craze. But it is more than a fad. It is an art form of the future with lasting value for the collector and connoisseur.

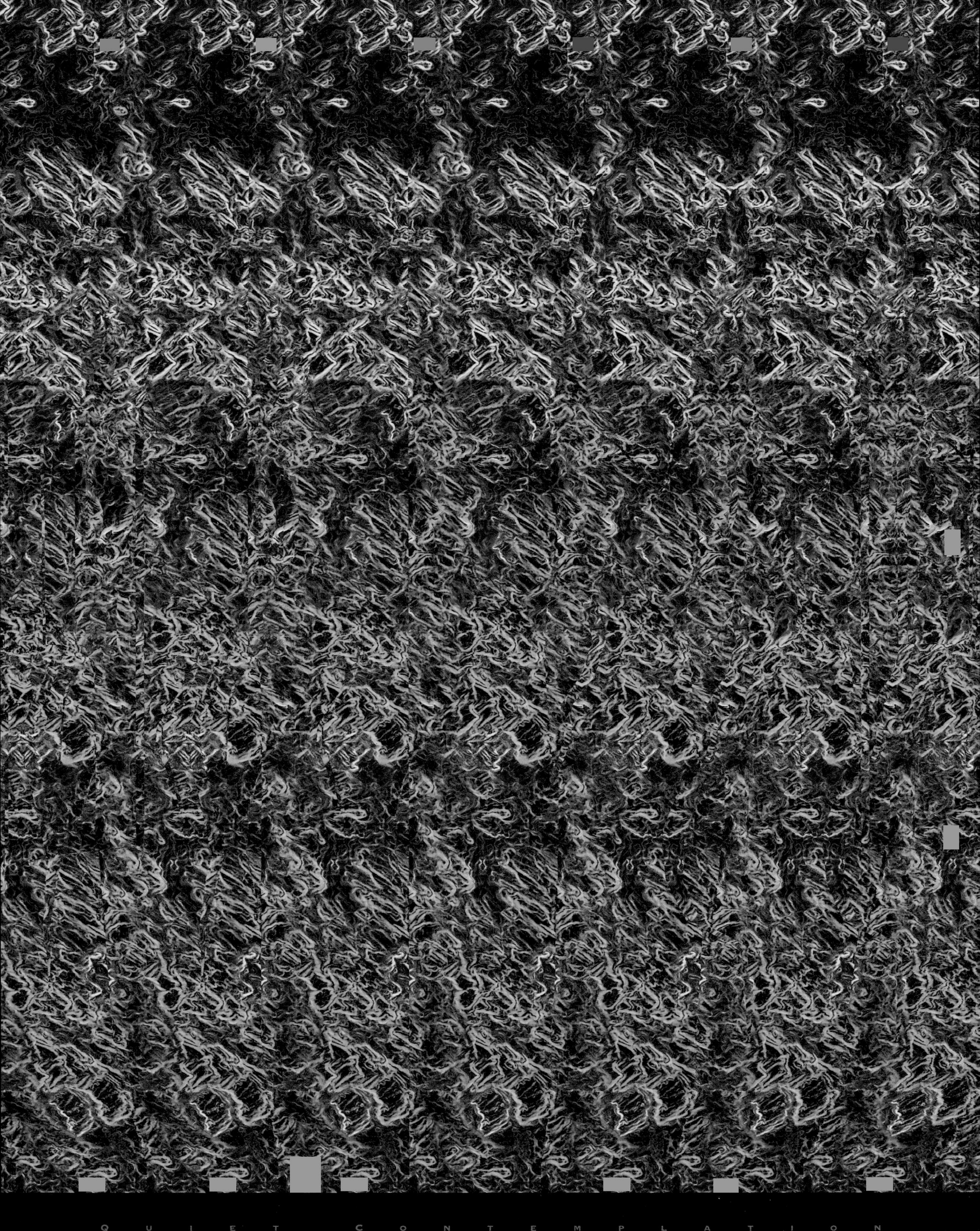

QUIET CONTEMPLATION

S H I M M E R I

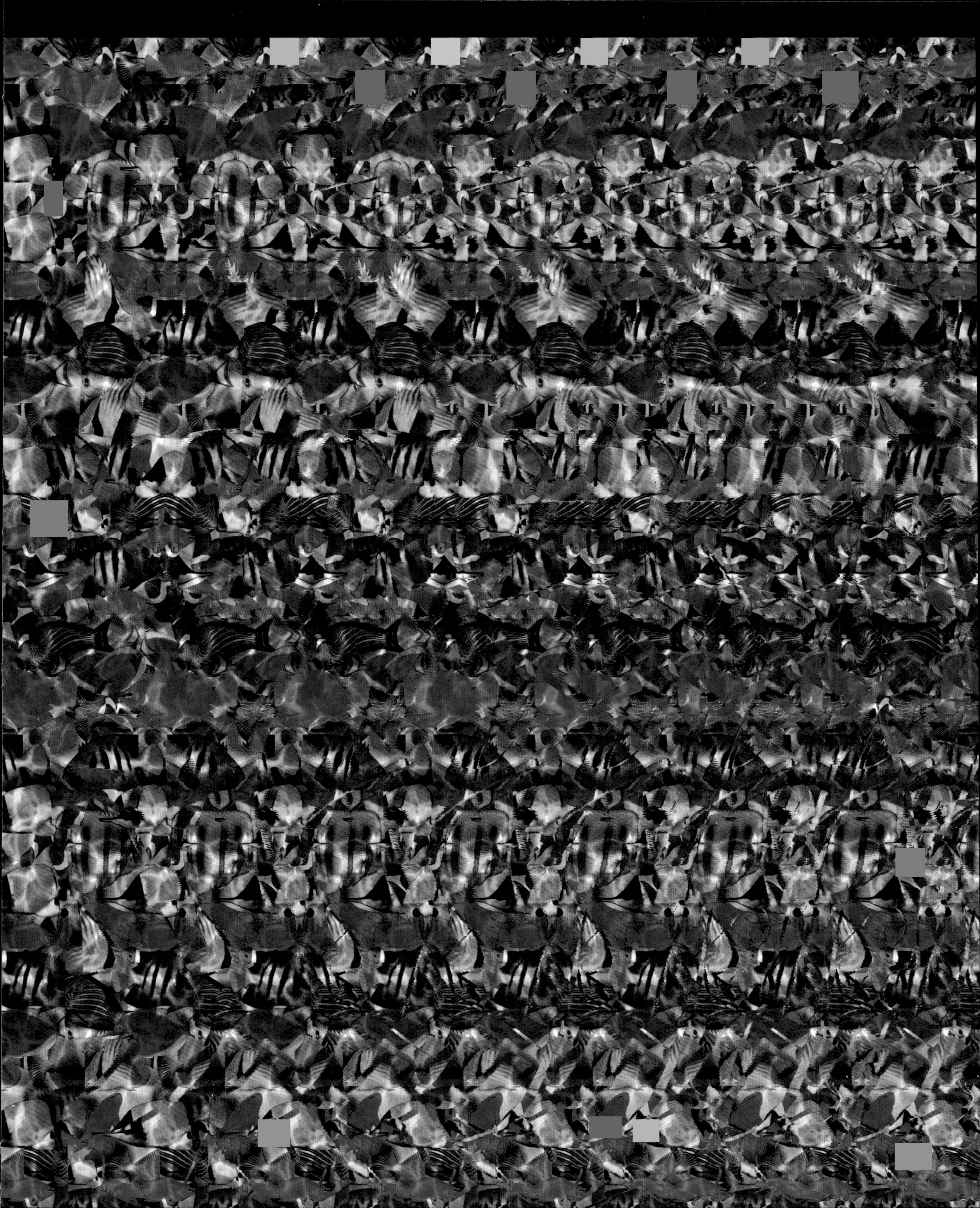

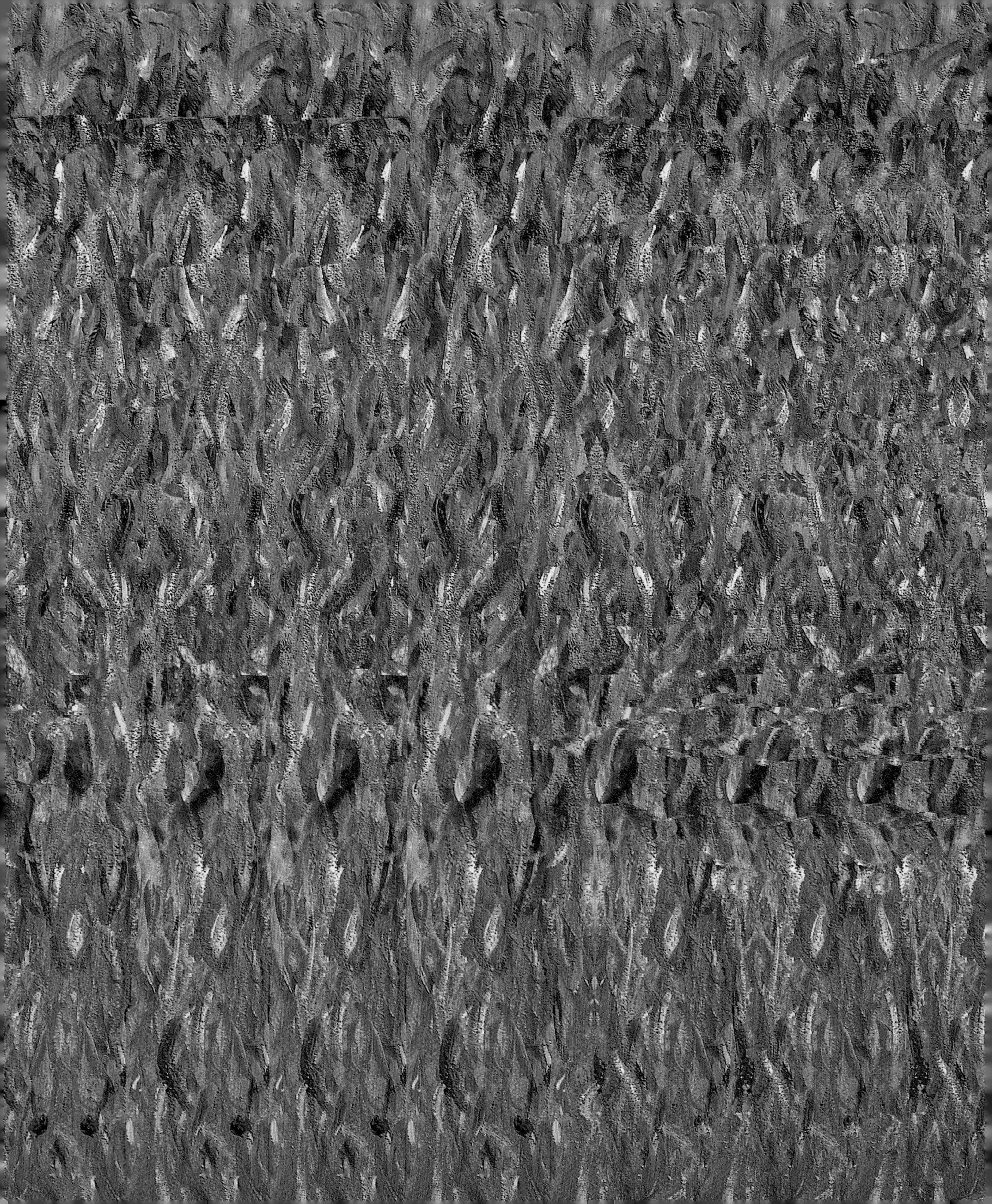

There is only one requirement to see Holusion Art—good vision in both eyes. A simple test will determine if you should be able to see the art. Hold your thumb up a foot or two in front of your eyes and stare across the room at the wall. If you see two thumbs, you are capable of stereo vision and should have no problem seeing the art. If you see only one thumb, then your stereo vision is lacking or poor, and your eyesight may inhibit your ability.

People who are nearsighted or farsighted should be able to see the art, as long as their vision is corrected with glasses or contacts. However, severe astigmatism makes the art much harder to see. A common misconception is that color-blind people are unable to see Holusion Art. Not true. It's the ***contrast*** of the pattern that hides the 3-D image, not the ***colors.***

Several optometrists and eye doctors have agreed that viewing Holusion Art will not hurt your eyes in any way. In fact, some vision therapists use focusing exercises much like those used in seeing the 3-D images.

One other tip: Keep the book level as you gaze into each image. If you rotate the book too far to the side, you can't see the images. However, you can turn the book upside down and still see them.

The reason the majority of non-seers have trouble with Holusion Art is purely psychological, resulting from a type of mental block. Barring the eye problems previously mentioned, just about everybody should be able to see the art with proper coaching and a little relaxation. A good coach can achieve success rates of 95% and up.

Holusion Art, if you think about it, is a very democratic, equal-opportunity art form. The ability to see these hidden images is not related to your IQ, age, maturity level or status in life. Young children can often see the images almost instantly, while their parents may struggle. So perhaps it helps to have a fresh mind open to new experiences.

In one way, Holusion Art is similar to the Rorschach test, those famous ink blots that supposedly reveal something about the viewer's inner life. People who are highly competitive, analytical or under stress often find it difficult to relax long enough to see the hidden image. As an experiment, you might try sending Holusion Art to someone who's usually dealing with high stress–a high-level executive, for instance. The exec may take a couple of days to find the image.

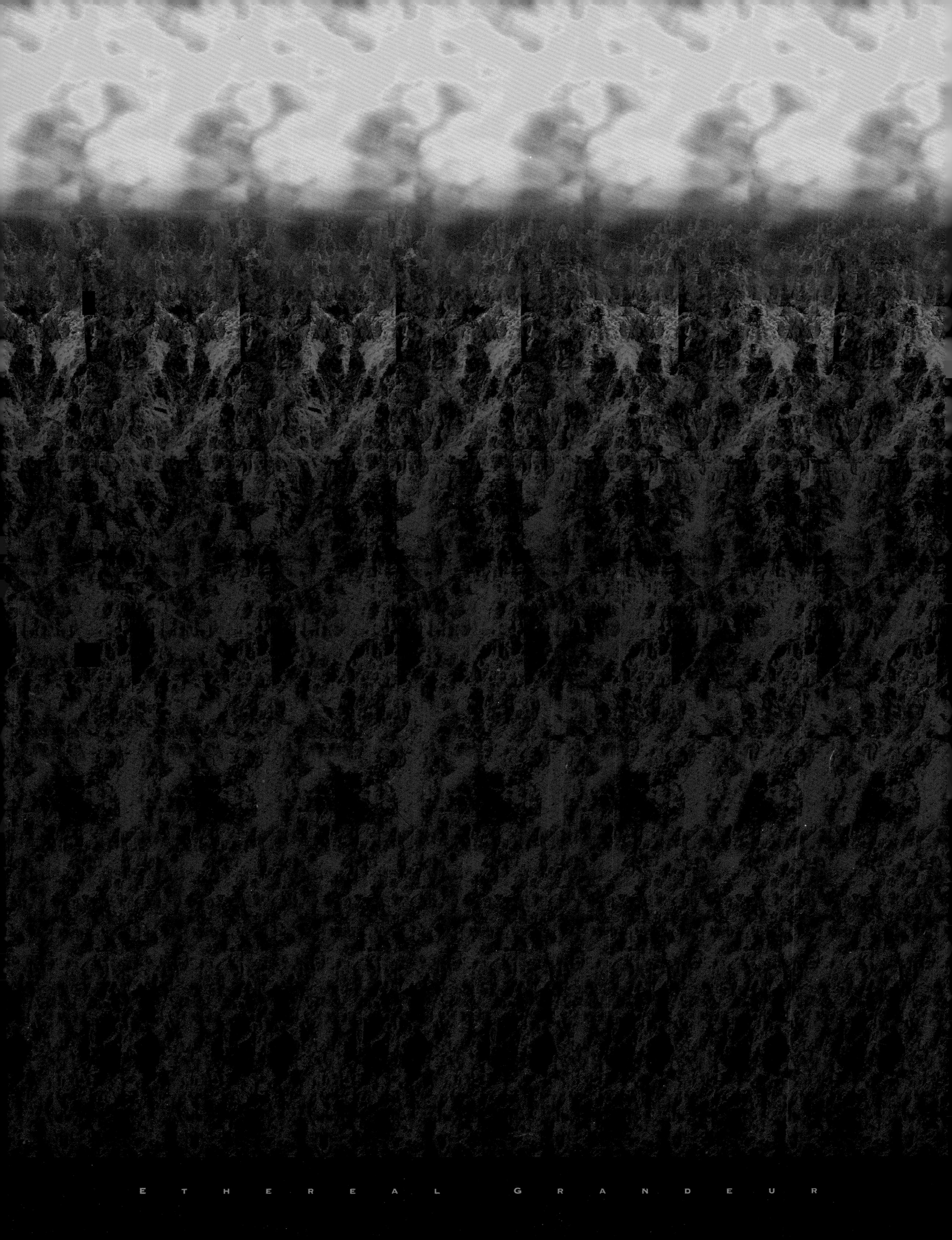
E T H E R E A L   G R A N D E U R

PINNACLE BEINGS

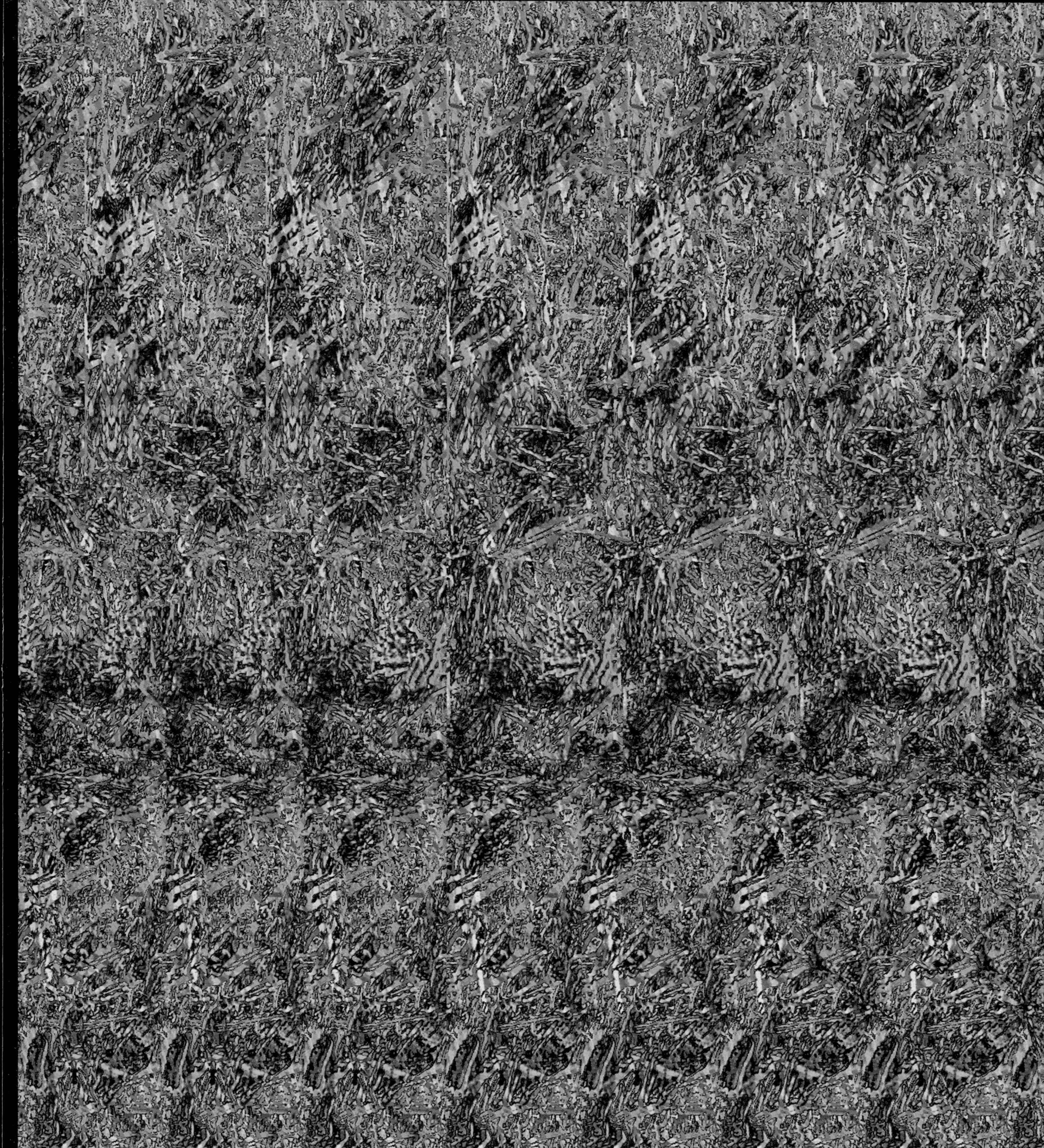

Because your eyes are spaced about two and a half inches apart, you see a slightly different perspective of the world with each eye. Scientists call this difference in perspective ***binocular disparity***, and it plays an important role in depth perception. Binocular disparity is the fundamental principle behind Holusion Art. Try the Sausage Trick to demonstrate binocular disparity. Touch the tips of your index fingers together, hold your hands a foot in front of your face and stare at a wall across the room. If you have good stereo vision, you should see the illusion of a small sausage pressed between your fingers.

Here's another intriguing optical illusion. Take a loose piece of paper in your right hand, roll it into a tube and look through it with your right eye. At the same time, place your left hand about six inches in front of your left eye, with the palm facing you, and the edge of your left hand resting against the tube. With both eyes open, stare across the room to see a hole in your hand.

These tricks work very much like Holusion Art. By looking past an object, you create an optical illusion because of binocular disparity.

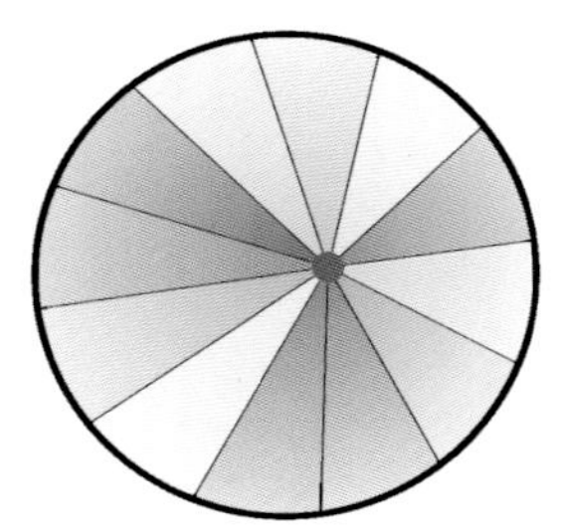
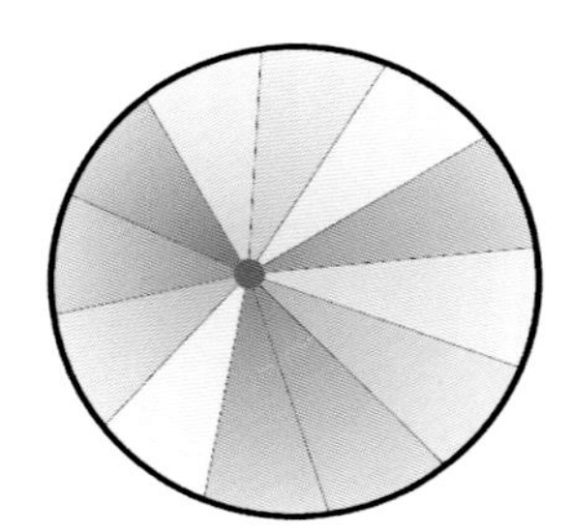

In 1838, Charles Wheatstone presented the first modern scientific explanation of binocular disparity and stereoscopic effects to the Royal Society of London. In his paper, Wheatstone presented several pairs of drawings called stereo pairs. Within each stereo pair, the drawings contained a slightly different perspective of an object to mimic the different views you see with each eye.

The illustration above is a replica of a stereo pair presented by Wheatstone. As you would with Holusion Art, focus past the page or use the enclosed transparent sheet to look at your reflection. Keep looking until the two objects fuse together.

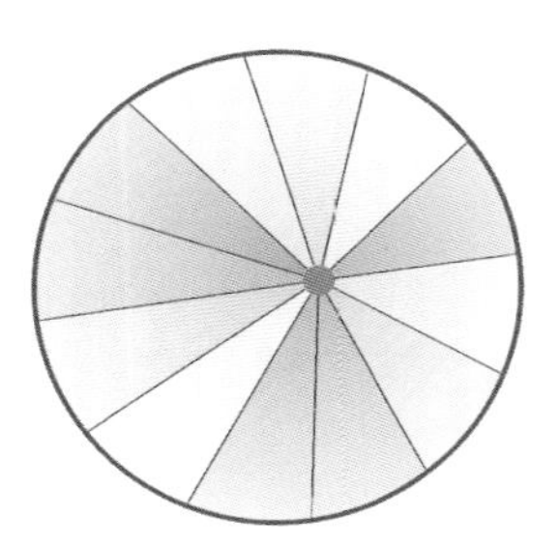
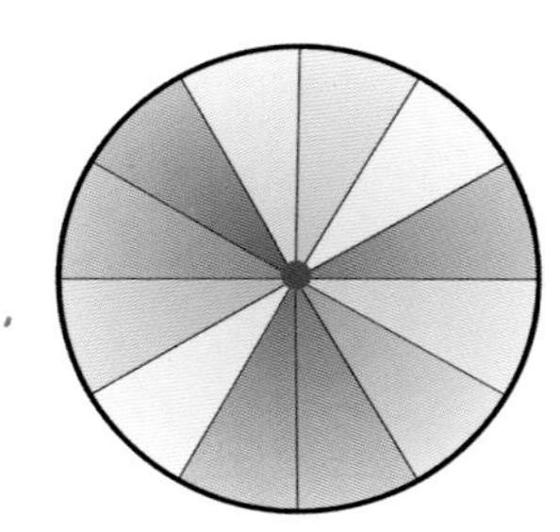
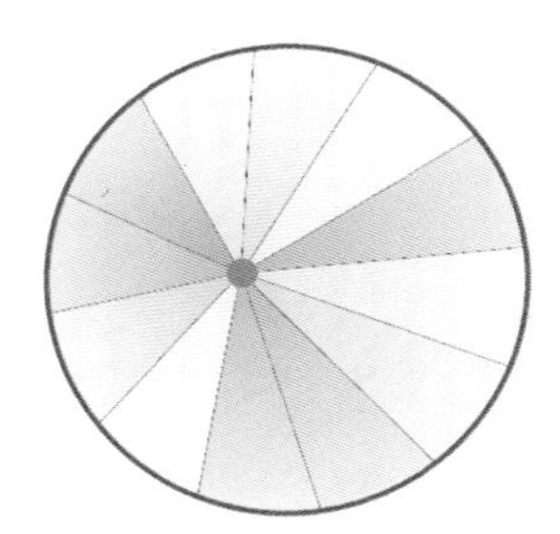

As the left object overlaps the right object, you see a strong image floating on the page, surrounded by a faint "ghost" image on each side, as demonstrated by the second illustration above. Because of the slight differences in perspective between the two objects, the strong central image appears to have depth. These stereo pairs rely on binocular disparity, as does Holusion Art, to achieve a 3-D illusion.

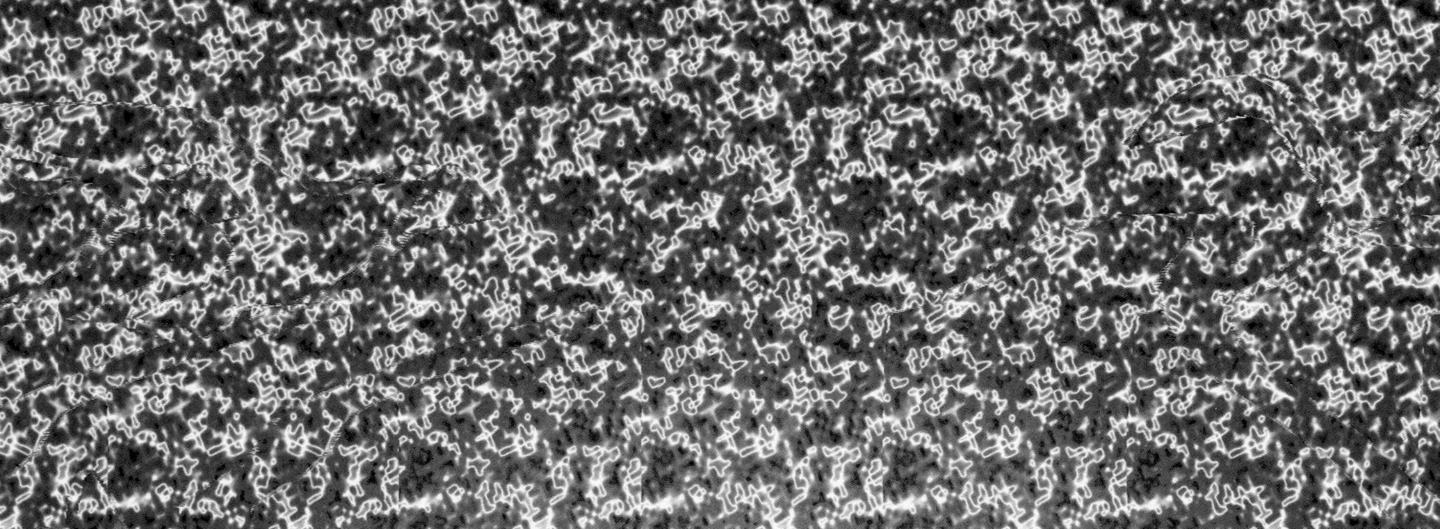

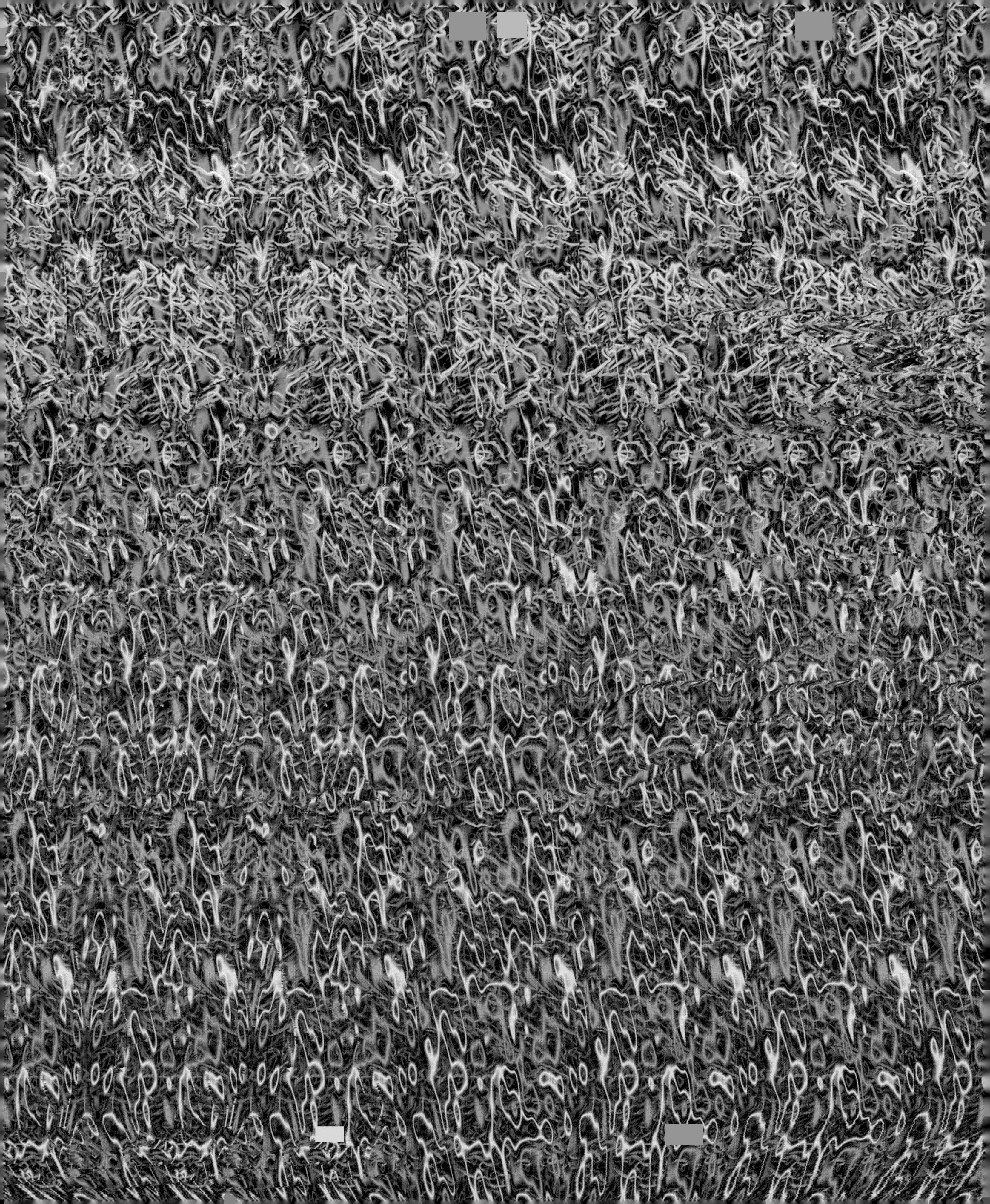

Holusion Art has evolved from a process developed and patented in the 1970's by a visionary artist named Donald Peck. Mr. Peck's work was done by hand, which was a very tedious process. In effect, he was like the scribes of long ago who copied biblical texts. In the 1990's, NVision automated and refined this process, bringing it to a new level through the use of the most advanced computers. Let's use a simple rose to demonstrate Mr. Peck's work.

Look at the roses above and focus past the page, just like you did with the stereo pairs. Use the transparent sheet if it helps. When your focus is just right, the first rose overlaps the second, and the second rose overlaps the third, and so on. This makes the roses appear to float on the page.

Because the eight roses above are spaced an equal distance apart, they appear to float on the same level. Mr. Peck discovered that he could change the perceived depth of any object by shifting it slightly to the left or right. The more an object is shifted to the left, the closer it appears to the viewer. The more it is shifted to the right, the farther it appears from the viewer.

Here, the third and fifth roses have been shifted slightly to the left, and the seventh one has been shifted slightly to the right. Notice the changes in depth as you focus through the page.

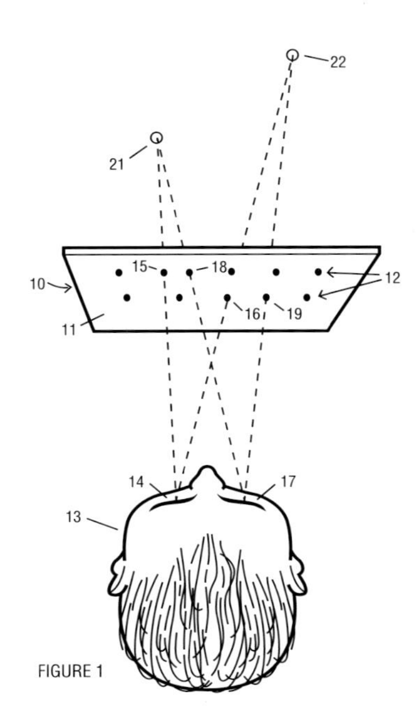

FIGURE 1

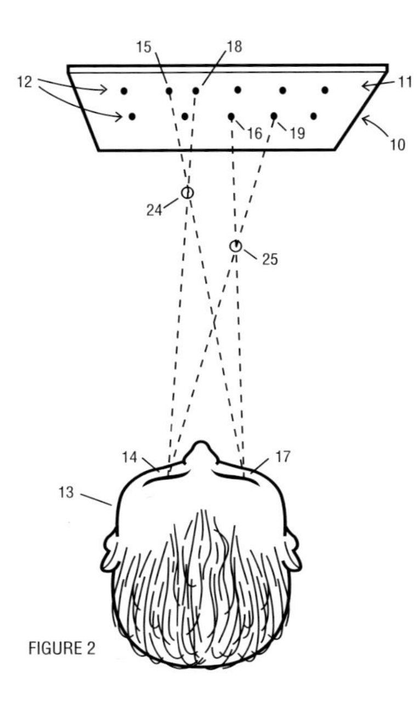

FIGURE 2

Here are *Figures 1* and *2* as originally depicted in the patent, which describe this phenomenon of repeating objects. In *Figure 1,* when you are focusing past the page, dot 15 as seen by your left eye merges with dot 18 as seen by your right eye to form the image of dot 21. The same applies to dots 16 and 19 as they merge to form the image of dot 22. However, since these dots are spaced farther apart, they appear to be at a greater distance from you.

Now look at the repeating roses again at the bottom of page 21. Rather than diverging your eyes and focusing through the page, you can cross your eyes and focus in front of the page. When you do this, you invert the stereoscopic effect, as shown by the images of dots 24 and 25 in *Figure 2*. This inverted viewing technique is difficult at first, and usually requires practice. The technique works with Holusion Art and can give some interesting and entertaining results. Objects that normally seem to come out at you are seen in "negative relief" and look like gelatin molds.

Designing images for inverted viewing is a new concept in Holusion Art—one of several new twists you can look for in the future. To view the image across the page, place your thumb between your eyes and the Holusion Art and focus on your thumb. Watch your thumb as you slowly move it closer and farther from your eyes, and at some point you should have the correct focus to see the inverted image.

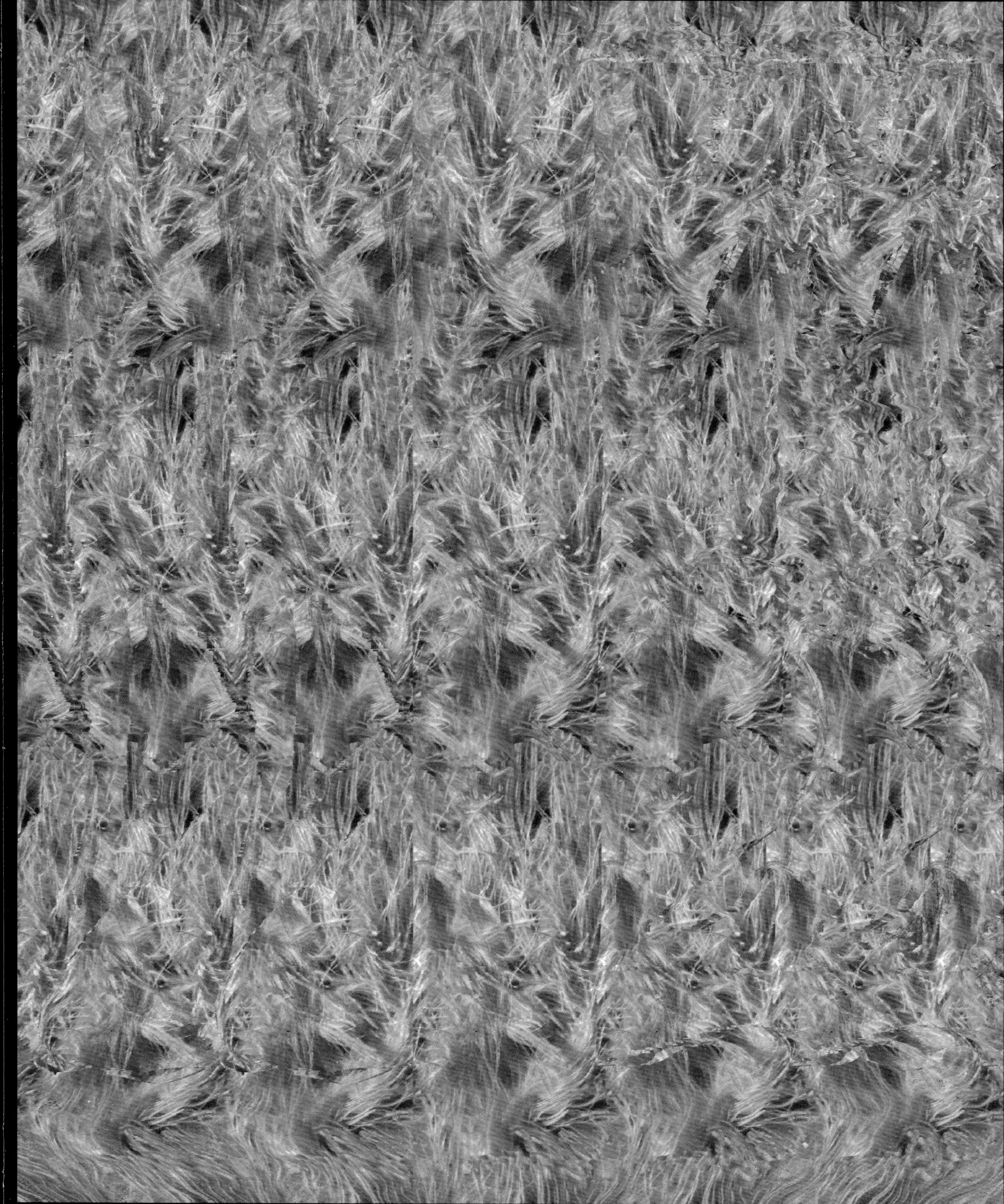

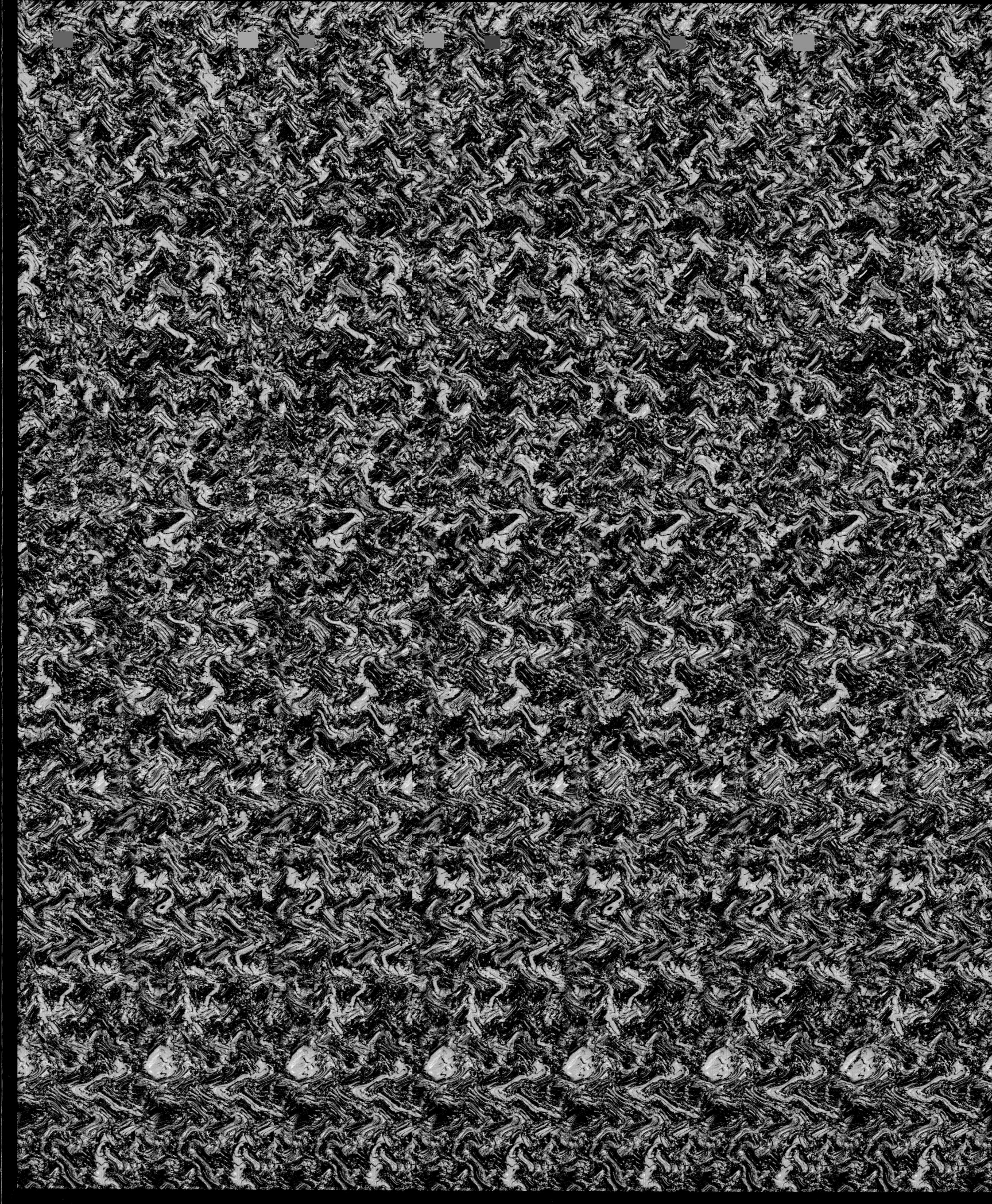

D O W N U N D E R

S U M M E R   W A L T Z

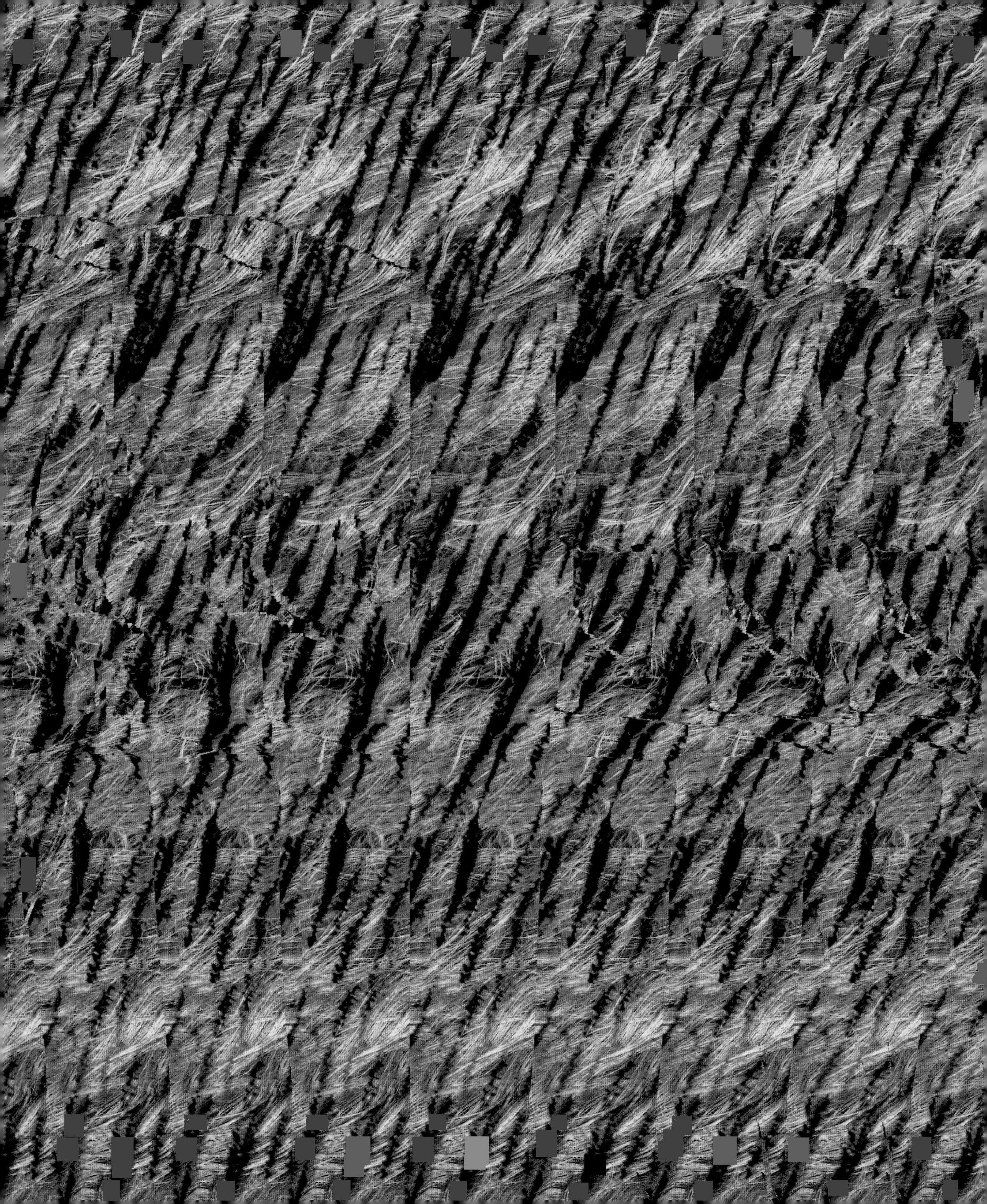

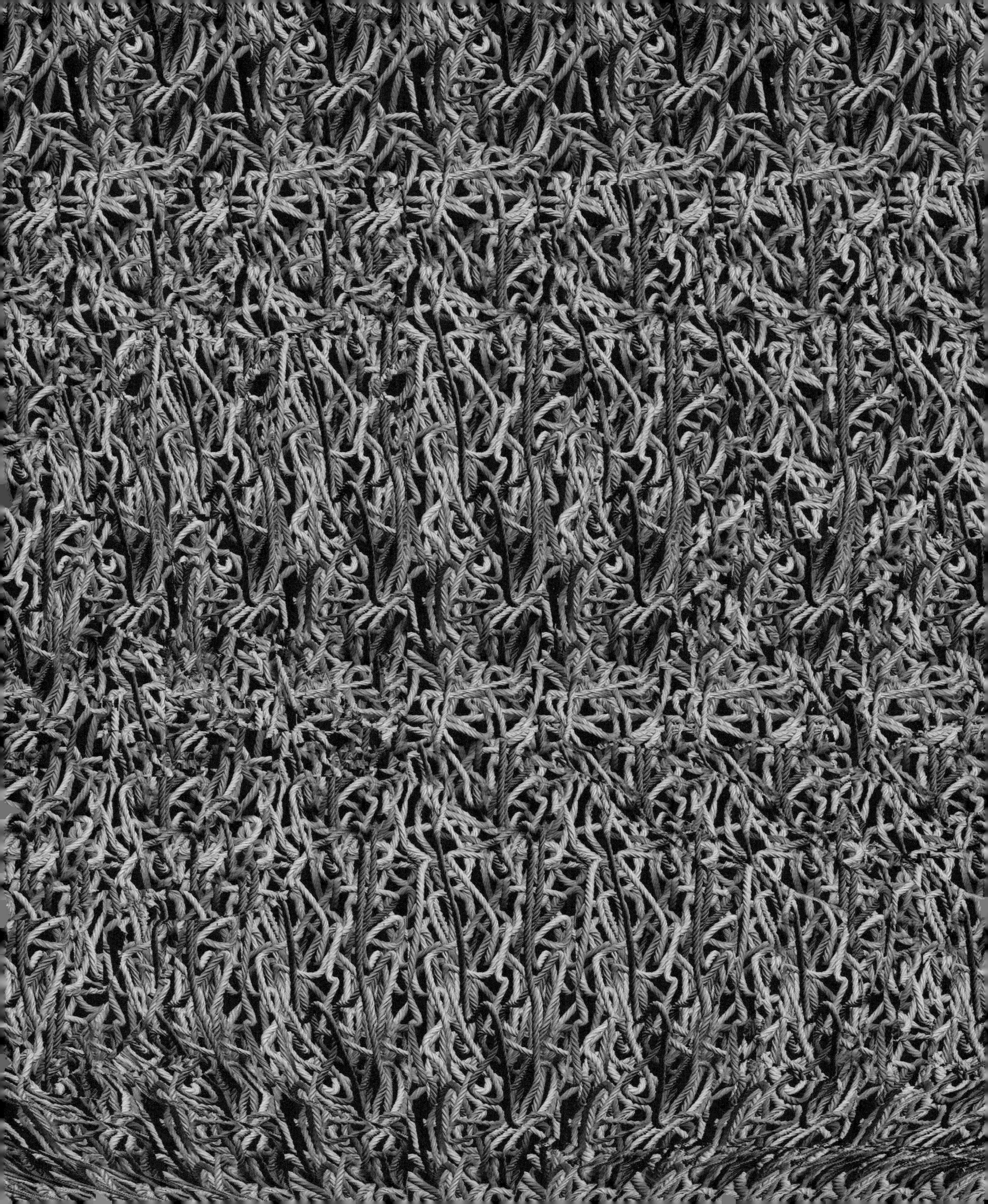

D R I V E

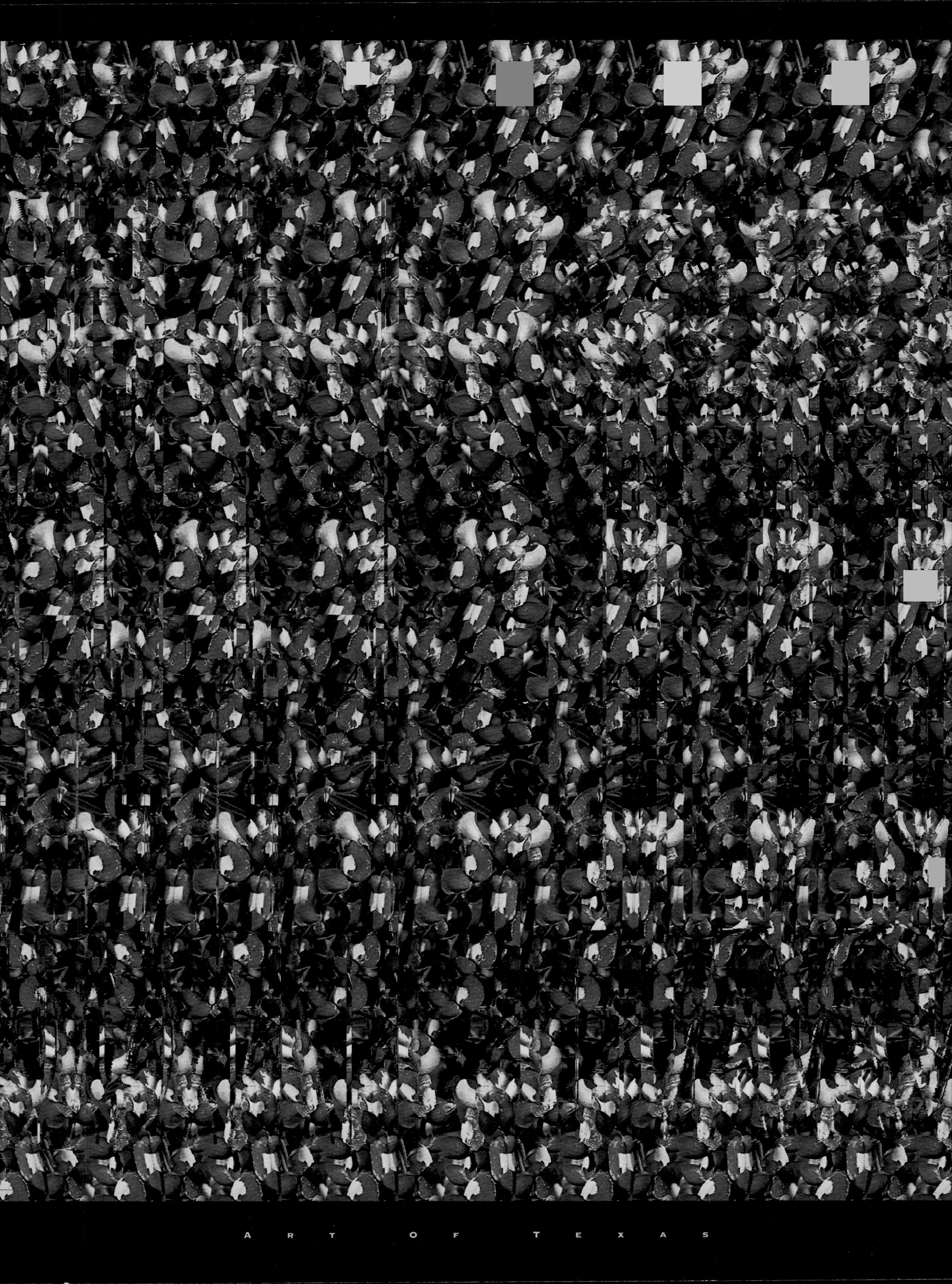
ART OF TEXAS

To bridge the conceptual gap between Mr. Peck's work and Holusion Art, imagine a row of a ***thousand*** roses, rather than eight, fitting in the same space. The roses would each have to be extremely small and would look more like dots than roses. But, no matter how small, each could be shifted left or right to create varying levels of depth.

Next, imagine a ***thousand*** rows, rather than just one, each with its own thousand repeating dots. This would create one million tiny dots filling a page. Each dot could be manipulated individually to achieve its own level of depth. The dots would act just like cells or atoms, each one insignificant and barely visible by itself. But when combined with neighboring dots, they could form complex 3-D images, such as dinosaurs, trains or anything else conceivable.

A 3-D print contains up to sixteen million repeating microscopic dots that are individually shifted to create the illusion of depth. The dots vary in color so that your eyes distinguish between them. The colors are artistically chosen to form the unique patterns you see on the surface of the print. Think of it as computerized pointillism.

Individually placing and moving sixteen million tiny dots is no easy task, and certainly cannot be done by hand. The prints usually take weeks or months to create, using high-powered computers that are fast enough and accurate enough to place the millions of microscopic dots.

Not only is this art form changing the way we look at *art*, it's changing the way we look, period. Gazing at 3-D images brings delighted smiles and puzzled expressions to the faces of millions—as you've probably noticed at storefronts where crowds gather to stare in amazement at the latest prints.

At some shopping malls, crowds have had to be dispersed because the congestion in front of store windows was creating a hazard. In some cities, wild rumors have spread, with "reliable sources" claiming that the art was the work of CIA defectors and NASA scientists.

Well, now you know the story of Holusion Art. You've traced the curious history of stereoscopic discoveries, and you've mastered the viewing techniques people use to see the images.

Holusion Art has taken giant leaps since the first creation of the B2 Stealth Bomber featured on the following page. Today, the artistic merit of the work is higher, and the intricate details within the aesthetic patterns have grown ever more sophisticated. This evolving art will dazzle your senses again and again. We have a vision—and you won't believe your eyes!

Quiet Contemplation

Shimmering Abyss

Lasting Impression

Ethereal Grandeur

Pinnacle Beings

Gates Of Gold

Extreme

Wet And Wild

Iron Horse

Grace Under Pressure

Unquestionable Beauty

Heights Of Paris

Down Under

First Lady

Summer Waltz

Siberian Solitude

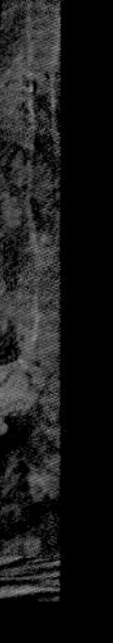